SCIENTIFIC AMERICAN EDUCATIONAL PUBLISHING

SCIENTIFIC AMERICAN INVESTIGATES SPIDERS

GIANT HUNTSMAN SPIDERS

BY ROXANNE ALDOLPHO

Published in 2025 by The Rosen Publishing Group
in association with Scientific American Educational Publishing
2544 Clinton Street, Buffalo NY 14224

Library of Congress Cataloging-in-Publication Data
Names: Aldolpho, Roxanne, author.
Title: Giant huntsman spiders / Roxanne Aldolpho.
Description: [Buffalo, NY] : Scientific American Educational Publishing, an imprint of Rosen Publishing, [2025] | Series: Scientific American investigates spiders | Includes index.
Identifiers: LCCN 2024006017 | ISBN 9781725351271 (library binding) | ISBN 9781725351264 (paperback) | ISBN 9781725351288 (ebook)
Subjects: LCSH: Heteropoda maxima–Juvenile literature. | Huntsman spiders–Juvenile literature. | Spiders–Juvenile literature.
Classification: LCC QL458.42.S63 A43 2025 | DDC 595.4/4–dc23/eng/20240320
LC record available at https://lccn.loc.gov/2024006017

Portions of this work were originally authored by Melissa Raé Shofner and published as *Giant Huntsman Spiders*. All new material in this edition is authored by Roxanne Aldolpho.

Designer: Andrea Davison-Bartolotta
Editor: Brianna Propis

Photo credits: Cover, p. 1 Chris Howes/Wild Places Photography/Alamy Stock Photo; p. 5 Tobias Hauke/Shutterstock.com; p. 7 pingebat/Shutterstock.com; p. 9 iSKYDANCER/Shutterstock.com; p. 11 Vinicius R. Souza/Shutterstock.com; p. 13 Nikhil Guhagarkar/Shutterstock.com; p. 15 EDPhotoZA/Shutterstock.com; p. 17 Jeroen Mikkers/Shutterstock.com; p. 19 Nick626/Shutterstock.com; p. 21 Barnaby Chambers/Shutterstock.com.

Printed in the United States of America

CPSIA compliance information: Batch #CSSA25. For Further Information contact Rosen Publishing at 1-800-237-9932.

CONTENTS

Words in the glossary appear in **bold** type the first time they are used in the text.

A GIANT DISCOVERY

Huntsman spiders are some of the largest and fastest spiders around! They were named after their hunting style because they catch **prey** themselves instead of using a web. Although they may look creepy, they're harmless to humans and even help with pest control.

A new species, or kind, of huntsman spider was discovered by a German scientist in 2001. This new species was much bigger than any other huntsman spider. It was named after its size: the giant huntsman.

Giant huntsmans are bigger and hairier than the many other kinds of huntsman spiders.

FUN FACT

THE AVERAGE LIFE SPAN FOR HUNTSMAN SPIDERS IS TYPICALLY AROUND TWO YEARS, BUT THEY CAN LIVE LONGER.

AROUND THE WORLD

Huntsman spiders can be found around the world in places with warm **climates**. Most species are native to Asia, but many kinds of huntsman spiders live in Australia. They're found in Southeast Asia, Europe, South Africa, South America, and New Zealand too.

Huntsman spiders also live in Hawaii and in states across the southern United States, including Florida, Texas, and Alabama, as well as California. The giant huntsman spider, however, has only been found in Laos, a Southeast Asian country bordering Thailand.

WHERE GIANT HUNTSMANS LIVE

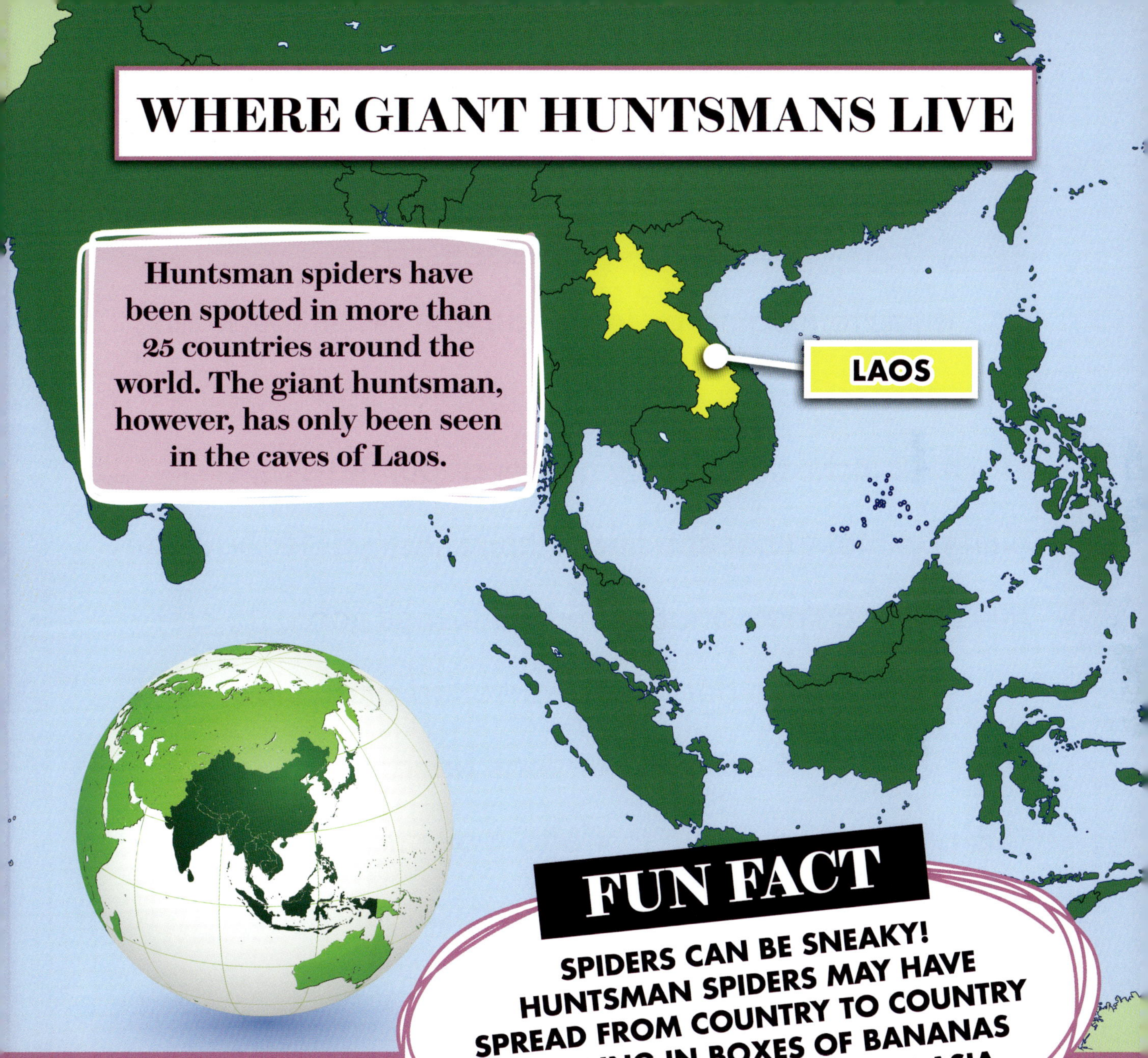

Huntsman spiders have been spotted in more than 25 countries around the world. The giant huntsman, however, has only been seen in the caves of Laos.

FUN FACT

SPIDERS CAN BE SNEAKY! HUNTSMAN SPIDERS MAY HAVE SPREAD FROM COUNTRY TO COUNTRY BY HIDING IN BOXES OF BANANAS THAT WERE SHIPPED OUT OF ASIA.

THE LARGEST SPIDER?

If you're judging by **leg span**, the giant huntsman is the largest spider in the world! Its leg span can reach 12 inches (30 cm), which is about the size of a dinner plate! Other species of huntsman spider have an average leg span of about 5 to 6 inches (13 to 15 cm).

Huntsman spiders are also called giant crab spiders. This is because their legs bend like crab legs. This allows them to run fast and move from side to side. It also helps them to climb straight, smooth surfaces such as glass.

If you notice a spider's legs bending out to the sides and forward—rather than upward and under the body—it's probably a huntsman spider. They have crab legs!

FUN FACT

ANOTHER HUGE SPIDER IS THE GOLIATH BIRD-EATING TARANTULA. BASED ON BODY MASS, IT'S CONSIDERED THE LARGEST SPIDER IN THE WORLD!

CAMOUFLAGING

Even though huntsman spiders are big, it's unlikely you'll spot them! Many species are brown or gray, which helps them blend in with their **environment**. Their legs may be banded, or striped.

Some huntsman spiders even have flat bodies that allow them to hide under loose tree bark, between rocks, and in small cracks in walls. Giant huntsman spiders look like other huntsman species, but they're lighter in color.

Huntsman spiders are excellent at camouflaging. This means they use their bodies to blend into their surroundings.

FUN FACT

BE CAREFUL IF YOU LIVE IN AUSTRALIA! SOME PEOPLE THERE HAVE FOUND HUNTSMAN SPIDERS IN THEIR CARS OR HOMES.

CARNIVORES

Most spiders are known for their webs, but huntsman spiders—including giant huntsman spiders—find food in another way. Instead, they hunt their prey. This is how they got their name! Huntsman spiders are carnivores, which means they eat meat. Their favorite meals include bugs, frogs, lizards, and even other spiders.

The giant huntsman spiders use their speed and **agility** to chase their prey. When they get close enough, they strike! They use their **venom** and sharp **fangs** to kill and eat their prey!

A giant huntsman's long legs let them move at speeds of up to 3 feet (0.9 m) per second!

FUN FACT

HUNTSMAN SPIDERS EAT A VARIETY OF SMALL CREATURES. SOMETIMES THEY'LL EVEN EAT MICE!

PROTECTING THE BABIES

Huntsman spiders don't spin webs, but their bodies still produce silk. A mother giant huntsman will make a sac from silk and lay up to 200 eggs inside. She'll hide the sac under a piece of bark or a rock. Some species carry their egg sacs while they move around.

It takes about three weeks for the eggs to **hatch**. During this time, a mother huntsman guards her egg sac if she doesn't carry it. She won't even leave to eat! Mothers protect their babies, called spiderlings, after they hatch too.

A mother huntsman spider will act **aggressively** to keep her babies safe.

FUN FACT

BABY HUNTSMAN SPIDERS MOLT. THIS IS WHEN THEIR SKIN BECOMES TOO SMALL AS THEY GROW AND IT PEELS OFF.

SPIDER FRIENDS

Unlike most other spiders, giant huntsman spiders are very social. *Delena cancerides*, one species of giant huntsman spiders, live in large groups called colonies. These spiders are sometimes called social huntsman spiders. They even share their food!

Social huntsman spiders are native to Australia. They build their colonies under the bark of dead trees. Several hundred spiders may live and work together. The mother spider keeps the colony safe as the other spiders leave it to hunt.

Many spiders are cannibalistic, meaning they eat their own kind. If food is short, giant huntsman spiders will eat each other.

FUN FACT

FEW OF THE 50,000 SPECIES OF SPIDERS ON EARTH ARE KNOWN TO LIVE IN GROUPS. SCIENTISTS ARE STILL FIGURING OUT WHY *DELENA CANCERIDES* LIVE TOGETHER IN COLONIES.

WHAT HAPPENS IF YOU'RE BITTEN?

Giant huntsman spiders live far away and are only found in caves, so it's unlikely a person would be bitten by one! But other types of huntsman spiders sometimes find their way into people's homes and cars. Luckily, most aren't usually aggressive. They may bite if they're bothered too much, but don't worry—it only hurts a little.

Some people may feel sick, throw up, or get a headache after being bitten by a huntsman. However, pain and swelling are often the only side effects. An ice pack should help, but make sure to see a doctor if it hurts for a long time.

How would a spider crawling up your arm make you feel?

FUN FACT

ALTHOUGH THEIR BITE IS VENOMOUS, A GIANT HUNTSMAN CAN'T KILL YOU!

PREDATOR AND PREY

Sometimes in the wild, the hunter becomes the hunted. Birds, spider wasps, **parasitic** worms, and lizards such as geckos find huntsman spiders to be tasty snacks. A huntsman's eggs may even be used as nests for babies of certain wasps and flies.

Try to stay calm if you come across a giant huntsman spider. They may look scary, but these creatures don't want to hurt you! If you find a huntsman in your house, an adult may be able to help you carefully catch it and take it outside.

Huntsman spiders might look creepy, but they play an important role in keeping the number of pests down.

FUN FACT

HUNTSMAN SPIDERS ARE GREAT AT CATCHING ANNOYING BUGS LIKE COCKROACHES!

GLOSSARY

aggressive: Ready to fight.

agility: The ability to move around quickly and easily.

climate: The average weather conditions of a place over a period of time.

environment: All the living things and conditions of a place.

fang: A hard, sharp-pointed body part a spider uses to put venom into its prey.

hatch: To break open and let young out.

leg span: The distance from the tip of one leg to the tip of the leg directly across from it.

mass: The total amount of matter in something.

parasitic: Having to do with a living thing that lives in, on, or with another living thing and harms it.

prey: An animal that is hunted by other animals for food.

venom: Something an animal makes in its body that can harm other animals.

FOR MORE INFORMATION

Books

Higgins, Melissa. *Splendid Spiders.* North Mankato, MN: Pebble, 2020.

Lundgren, Julie K. *Spiders.* New York, NY: Crabtree Publishing Company, 2022.

Websites

Australian Geographic
https://www.australiangeographic.com.au/topics/wildlife/2021/07/australian-huntsman-spiders-your-friendly-neighbourhood-insect-control/
Many common questions about the huntsman are answered here.

FirstAidPro
https://www.firstaidpro.com.au/blog/huntsman-spider-bite-how-dangerous
Learn more about huntsman spiders—and what to do if you're ever bitten by one!

Publisher's note to educators and parents: Our editors have carefully reviewed these websites to ensure that they are suitable for students. Many websites change frequently, however, and we cannot guarantee that a site's future contents will continue to meet our high standards of quality and educational value. Be advised that students should be closely supervised whenever they access the internet.

INDEX